KNOW HOW SCIENCE

FIRST EDITION

BY PRASUN BARUA

ABOUT

Welcome to KNOW HOW SCIENCE! This is a nonfiction science book which contains various types of articles on how-to related science topics. Science is a process that helps us understand and explain the world around us. It significantly contributes to the production of knowledge. There is a significant contribution of science in the development of modern civilization. It made our life easy and comfortable. Life changing contribution of science is significant for us. This book covers various types of articles categorized energy, power plant, electrical and electronic device. Thanks for reading the book.

CONTENTS

CHAPTER NO.	TITLE	PAGE NO.
CHAPTER-1	HOW DOES A NUCLEAR POWER PLANT WORK?	5
CHAPTER-2	HOW DOES AN ELECTRIC MOTOR WORK?	8
CHAPTER-3	WHAT IS SERVO MOTOR AND HOW IT WORKS?	13
CHAPTER-4	HOW DOES A HYDROGEN FUEL CELL WORK?	17
CHAPTER-5	WHAT IS LIQUEFIED PETROLEUM GAS AND HOW IT IS PRODUCED?	20
CHAPTER-6	HOW HYDROELECTRIC POWER PLANT WORKS?	23
CHAPTER-7	HOW ELECTRIC CARS WORK?	26
CHAPTER-8	WHAT IS MICROCONTROLLER AND HOW IT WORKS?	30
CHAPTER-9	WHAT IS MICROPROCESSOR AND HOW IT WORKS?	34
CHAPTER-10	WHAT IS BIOMASS AND HOW TO CONVERT BIOMASS INTO ELECTRICITY?	40
CHAPTER-11	HOW SOLAR CELLS WORK?	44
CHAPTER-12	HOW SOLAR PANELS ARE MANUFACTURED?	46
CHAPTER-13	HOW DIODES WORK IN A SOLAR PANEL?	50
CHAPTER-14	WHAT IS OCEAN ENERGY AND HOW DOES IT WORK?	52
CHAPTER-15	WHAT IS GEOTHERMAL ENERGY AND HOW IS IT USED TO GENERATE ELECTRICITY?	54
CHAPTER-16	WHAT IS WIND ENERGY, HOW DOES WIND POWER WORK AND HOW TO DESIGN A WIND FARM?	57

CHAPTER-17	HOW DOES ARTIFICIAL INTELLIGENCE WORK?	60
CHAPTER-18	WHAT IS BIOELECTRICITY IN SCIENCE?	64
CHAPTER-19	WHAT IS ELECTRICITY IN SCIENCE?	66
CHAPTER-20	HOW DOES A SOLAR THERMAL POWER PLANT WORK?	68
CHAPTER-21	WHAT IS GREEN HYDROGEN AND HOW IS IT PRODUCED?	71

CHAPTER-1
HOW DOES A NUCLEAR POWER PLANT WORK?

A power plant that produces power from atomic energy, delivered as a nuclear power through an atomic splitting chain response inside the vessel of an atomic reactor is called nuclear power plant. The primary part of a nuclear power plant is the atomic reactor, which contains the atomic fuel (generally uranium) and has frameworks that make it conceivable to begin, maintain and stop the atomic response in a controlled way.

How does a nuclear power plant work?

How a nuclear power plant works is like an ordinary warm plant, where nuclear power is gotten through the ignition of petroleum products. In an atomic reactor, nonetheless, this energy is acquired through the atomic splitting chain responses of the uranium molecules from the atomic fuel.

A nuclear power plant generates electricity from the nuclear power delivered by atomic parting chain responses in the vessel of an atomic reactor. The freed nuclear power is utilized to warm water at high tension and high temperature until it becomes steam. This steam turns a turbine associated with a generator which converts the mechanical energy of the turbine into electrical energy.

There are various kinds of atomic reactors, however, two extraordinary plans are available in more than 80% of the right around 450 usable units on the planet:

- Compressed Water Reactor (PWR)
- Bubbling Water Reactor (BWR)

How does a PWR reactor function?

It is essential to remember that in atomic splitting the cores of weighty molecules are assaulted with neutrons and afterward deteriorate into more modest and lighter cores. Whenever this happens they discharge the energy that ties the neutrons and protons that make them, and afterward, they transmit a few neutrons. These can deliver more splitting as they collaborate with new weighty cores, which will then radiate new neutrons, etc., so the response supports itself. This duplicating impact is known as the atomic parting chain response.

How a nuclear power plant works can be streamlined into five phases?

1. The uranium splitting happens inside the atomic reactor. It delivers a lot of energy that warms the coolant water circling at extremely high tension. This water is moved using the essential circuit to a heat ex-changer (steam generator) that produces water steam.
2. This steam is moved to the generator-turbine set using an optional circuit.
3. When there, the vanes in the turbine move the alternator and the technician energy is changed into power.
4. Whenever the water steam goes through the turbine it is shipped off a condenser to chill and become fluid water once more.
5. The water is then moved to the new steam generator to become steam again inside a shut circuit.

Primary parts

It was recently demonstrated that an atomic reactor is a site that can start, support and stop atomic splitting chain responses in a controlled manner, with sufficient means to extricate the produced heat. The focal part of a nuclear power plant is the reactor, the site that houses the atomic fuel. Its principal parts are:

- ➤ **Fuel:** The material, as a rule, advanced uranium dioxide, where the parting responses occur. It is utilized all the while as a wellspring of energy and neutrons to support the chain response. It is introduced in a strong state as barrel-shaped pills typified into metallic poles a couple of meters long.
- ➤ **Mediator:** Water that dials back the quick neutrons produced by the splitting, which prompts new partings and the sustenance of the chain response.
- ➤ **Cooling Water:** The very water that prompts the splitting as a mediator currently extricates the hotness created by the parting response from the uranium in the fuel.
- ➤ **Control Bars:** The control components in the reactor. They go about as neutron safeguards. These bars are made of indium-cadmium or boron carbide and make it conceivable to continually control the neutron populace while keeping the reactor stable; they likewise make it conceivable to stop the response at whatever point vital.
- ➤ **Protecting:** It forestalls the hole of radiations and neutrons from inside the reactor to the outside. Normally safeguarding is comprised of cement, steel or lead.
- ➤ **Security components:** All nuclear power plants have numerous well-being frameworks to forestall the hole of radioactivity to the outside. These frameworks incorporate the regulatory structure.

CHAPTER-2
HOW DOES AN ELECTRIC MOTOR WORK?

What is an electric motor?

Electric motor is an electrical machine that changes energy from one form to another. It changes the electrical energy into a mechanical one. Electric motor is a kind of pivoting gadget. It works based on the theory of electromagnetism. It collaborates attractive field of the motor. The attractive field connects with the electric flow in the winding wires. This communication produces force as force. This force is applied to the shaft of the motor. Direct flow or alternating flow is utilized to control the electric motor. Direct current is moved by batteries or rectifiers. Exchanging flow is moved by inverters, electrical generators, and power matrices. Electric motors are characterized by many variables. Like the kind of force source, applications, and so forth. Electric motor is utilized in blowers, machine devices, power apparatuses, siphons, fans, and turbines etc.

Guidelines for an electric motor

Each instrument has an alternate guideline. The rule depicts the hypothesis on which the instrument works. The electric motor likewise has a characterized standard.

An electric motor deals with the rule that when a flow is passed through a rectangular loop put in an attractive field, power is applied to the curl. This power is responsible for the nonstop revolution of the motor. In light of this revolution, energy transformation happens. In other words, the guideline of the electric motor is transferred to a flow conveying conduit. This current-conveying guide delivers an attractive field. This current-conveying guide is set in the opposite direction of the attractive field. Because of this, it encounters power.

Development of an electric motor

Each gadget has had an exceptional development. It is important to understand the developments. Here is a clarification of the development of an electric motor.

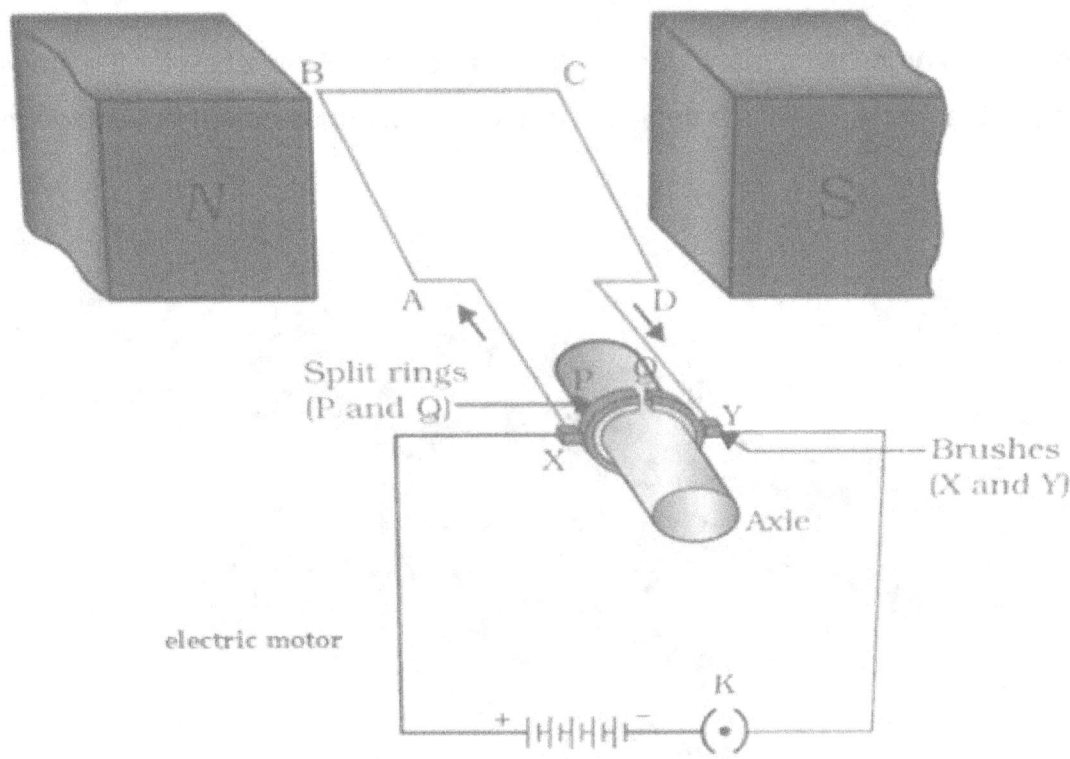

- It has a rectangular loop of wire ABCD.
- It has a solid horseshoe magnet. The curl ABCD is placed opposite to this magnet.
- The closures of the curl ABCD are associated with split rings P and Q. These split rings assume the part of the commutator. It helps invert the course of the ongoing stream.
- The innermost portion of the split rings is protected. It is connected to the hub. The pivot is allowed to turn.

> The outside side of the directing edges of the split rings is associated with the fixed brushes. These brushes, X and Y are associated with the battery. This finishes the circuit.

Portions of an electric motor

An electric motor involves many parts. These parts are fundamental for the smooth working of the motor. Here is a depiction of the significant pieces of an electric motor.

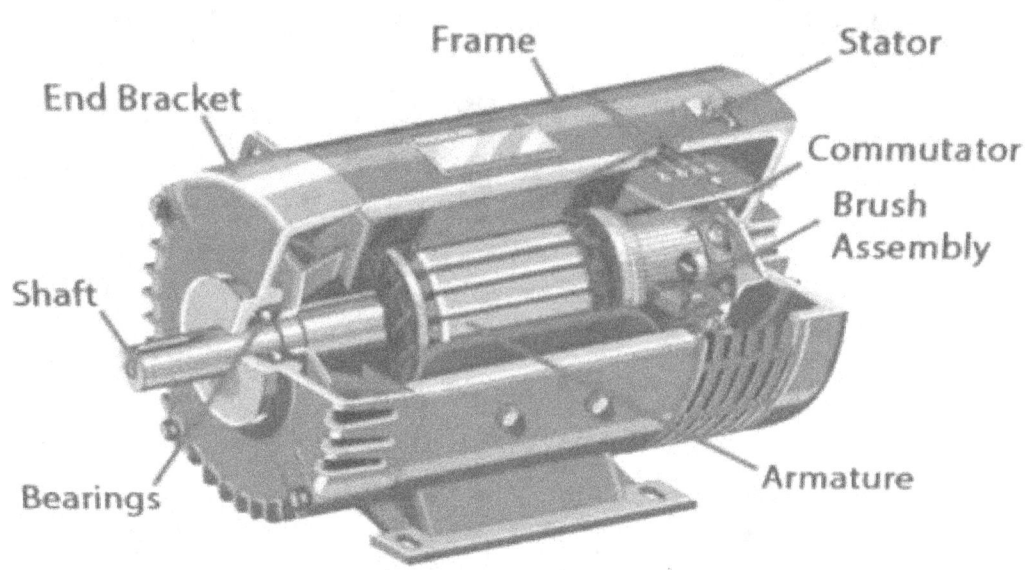

> **Rotor:** The rotor is a moving piece of the motor. Its job is to pivot the shaft of the motor. This revolution in the shaft produces mechanical power. The rotor likewise includes a guide. This guide conveys flows. It also aids in conversing with the attractive fields present in the stator.
> **Bearing:** Bearings are utilized to offer help the rotor. This is fundamental to enacting the pivot of the rotor. With the assistance of these, the shaft of the motor grows. It reaches out to the heap of the motor.

- **Stator:** This is an idle piece of the electromagnetic circuit of the motor. It contains extremely durable magnets and windings. The stator can be made of slim metal sheets. They are called covers. They assist in diminishing energy misfortune.
- **Windings:** Wires laid inside the loop of an electric motor are called "windings." This makes for magnet spasm posts when the current is provided.

The operation of an electric motor

The electric motor, as referenced, is a turning gadget. The working of an electric motor makes sense of its instrument. Here are a few stages which make sense of the workings of an electric motor.

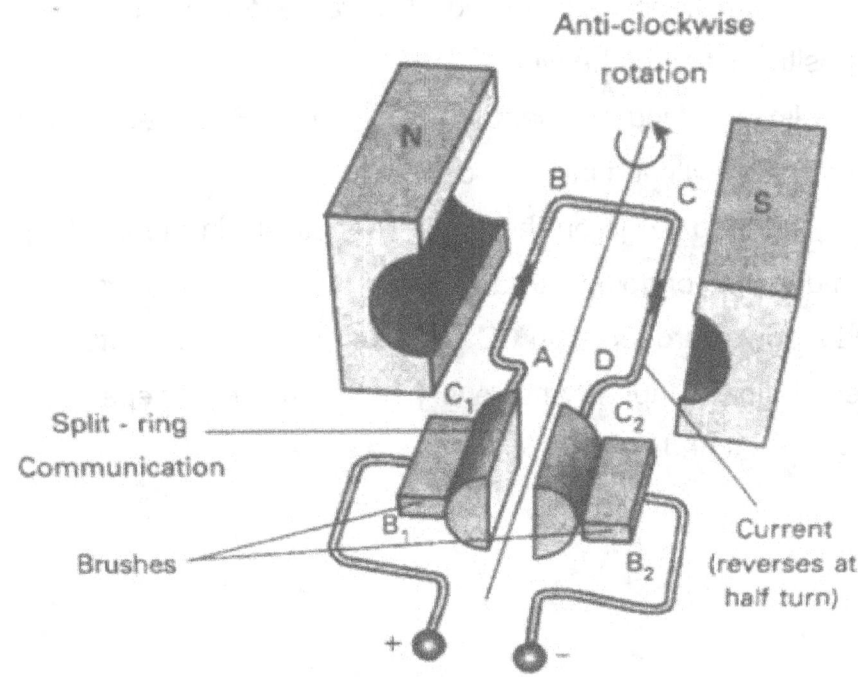

- When the battery of the motor is turned on, current streams into it. Current flows from A to B through loop AB. During this, the attractive field bearing is from North to South. With Fleming's Left-Hand Rule, a power acts downwards on AB. On CD, a vertical power is applied in this manner. Because of this, the loop turns. The abdominal muscles contract and the CD rises.
- Presently, the two curls, AB and CD, have exchanged positions. Presently, the progression of current is from C to D. Furthermore, the attractive field course is from North to South. Loop CD gets vertical power and it moves upwards. Loop AB moves downwards. So both the loops do half pivots.
- An electric motor requires a full turn to work. To acquire this, the bearing of the ongoing stream is changed. Utilizing a commutator, the heading of the current is changed. A commutator has two divided rings. Brushes are additionally connected to its circuit.
- As the pivot of the loop starts, the rings likewise turn. When the loop becomes attracted to the attractive field, the brushes contact the hole between the rings. Because of this, the circuit breaks.
- In light of inactivity, the ring keeps on moving. The far edge of the ring gets associated with the positive finish of the wire.
- Part rings P and Q are linked to independent loops CD and AB. Because of this, the heading of the current is switched in the circuit.
- Curl CD is on the left and loop AB is on the right. The current in the CD loop is switched at the moment. the ongoing streams from D to C. A vertical power follows up on AB and a downward force on CD. This keeps the loop pivots.
- This inversion of electric flow happens after every half turn. This keeps the curl pivoting till the battery is switched off.

CHAPTER-3
WHAT IS SERVO MOTOR AND HOW IT WORKS?

What is servo motor?

A closed-loop system which contains a servo motor, shaft, drive gears, control circuit, potentiometer, amplifier and encoder is called servo motor. It is an independent electrical device which can rotate and control machine's part smoothly with high efficiency. With a particular angle, position and velocity, it's output shaft can be rotated. A regular motor is used in this type of motor which is coupled with a sensor for positional feedback. In order to serve this purpose, a controller is designed for the motor. It can regulate the rotational or linear speed and position by integrating positional feedback which is the part of a closed-loop mechanism of the motor. It can regulate movements of the shaft with an electrical signal. Speed and position feedback are delivered by a sensor. In this case, a type of encoder works as a sensor.

Classifications of servo motor

Based on applications, there are various types of servo motors. They are: AC servo motor, and DC servo motor. A servo motor can be evaluated based on three factors. They are: type of current (whether it is DC or AC), motor brushes (commutation) and motor rotation (whether it is synchronous or asynchronous). Let's discuss the first servo consideration. AC or DC consideration is the most basic classification of a motor based on the type of current it will use.

DC and AC motors can be differentiated based on their ability to regulate the speed. At a constant load, the speed in a DC motor is directly proportional to the supply voltage with a constant load. On the other hand, speed in an AC motor is determined by the number of magnetic poles and the applied voltage's frequency. AC motors can resist high current. They are broadly used as robots where high accuracy and recurrences are required. For example, we can mention say in-line manufacturing and other industrial applications. A DC Servo Motor can be brushed or brush less. In electronically, it is brush less. Operation of Motors with brush are quite easy and comparatively cheap. On the other hand, brush less motors provide high efficiency with little noisy and are more dependable.

In order to reverse the current direction periodically between the rotor and the drive circuit, a rotary electrical switch is used which is known as a commutator. It contains a cylinder having multiple metal contact segments on the rotor. Multiple electrical contacts known as "brushes" built with a soft conductive material like carbon press against the commutator, creating a sliding contact with commutator's segments as it rotates. Most of servo motors are AC brush less. In order to reduce the cost and for easy operation, brushed permanent magnet motors work as servo motors sometimes. Permanent magnet DC motor is an example of typical type of brushed DC motor. An encoder or hall effect sensor is used in brush less DC motors for communicating electronically.

Rotating speed of the rotor is as like as the speed as the rotating magnetic field of the stator in a synchronous motor. On the other hand, in an asynchronous motor, also known as an induction motor, the rotor rotates with a lesser speed than rotating magnetic field of the stator. By changing the frequency and the number of poles, it's speed can be regulated.

How servo motor works?

DC servo motor operates based on four core components like a DC motor, a gear assembly, a control circuit and a position sensing device. By applying voltage, DC motor's required speed can be achieved. Voltage is produced by a potentiometer for controlling the speed of the motor. A control pulse can also be used in some circuits for producing DC reference voltage which is applied to a pulse width voltage converter. In order to generate the pulses in terms of duty cycles for getting accurate digital control, a PLC (Programmable Logic Controller) is used. A voltage corresponding to the absolute angle of the motor shaft through the gear mechanism can be produced by using the feedback signal sensor which is usually a potentiometer. Feedback voltage value is applied to the amplifier which compares the voltage produced from the current position of the motor resulting from the potentiometer feedback and to the required motor's position causing an error either of a positive or negative voltage. Error voltage is applied to motor's armature which is amplified by the amplifier for energizing the armature. Until error becomes zero, the motor rotates. The armature voltage reverses when the error is negative. As a result, the armature rotates in the opposite direction.

Two core components such as stator and rotor exist in asynchronous AC servo motor. There are two components exist in the stator. They are: cylindrical frame and stator core. Around the stator core, the armature coil is wound and the coil is connected to a lead wire through which current is supplied to the motor.

In brush less servo motor, a permanent magnet exists in the rotor which is induced by electromagnetism. Synchronous form is achieved when the rotor is synchronized with the energized field of the stator. If the stator field is de-energized, the rotor stops rotating. As there is no rotor current in these motors, the efficiency of these motors are very high. In order to give feedback to the servo motor controller, an encoder is added to the rotor.

Stator core, lead wire, armature winding, rotor with shaft and the rotor core winding exist in the asynchronous or induction AC servo motor. The rotor or squirrel cage is available in most induction motors. After getting AC supply, an alternating flux field is generated around the stator winding which resolves with synchronous speed. This revolving flux is also known as a rotating magnetic field (RMF).

According to electromagnetic induction law of Faraday, an induced electromagnetic force in the rotor conductors is produced due to the relative speed between stator rotating magnetic field and rotor conductors. An alternating flux field around the rotor is generated by the induced current in the rotor known as rotor flux which lags behind the stator flux.

The rotor velocity is related between the rotating stator flux field and the rotor rotates in the same direction as that of the stator flux. In this case, asynchronous type is originated as the rotor speed can't synchronize with the stator flux speed.

CHAPTER-4:
HOW DOES A HYDROGEN FUEL CELL WORK?

A hydrogen fuel cell utilizes the compound energy of hydrogen to deliver power. It is a perfect type of energy with power, intensity, and water as the main items and side-effects. Energy components offer an assortment of uses, from transportation to crisis back-up power, and can drive frameworks as extensive as a power plant or as small as a PC. Fuel cells give benefits over conventional burning-based advances, including more prominent efficiencies and lower discharges. Since hydrogen fuel cells just emanate water, there are no carbon dioxide outflows or different toxins delivered into the air. Power modules are additionally calm during activity, as they have fewer moving parts than ignition advances.

How does a hydrogen fuel cell work?

A hydrogen fuel cell produces power by utilizing a synthetic response. Each energy component has two terminals: a negative anode and a positive cathode. The response to delivering the power occurs at these anodes, with an electrolyte conveying electrically charged particles among them and an impetus to accelerate the responses.

Hydrogen is used as the fundamental fuel in a hydrogen energy unit. However, the cell additionally needs oxygen to work. Probably the biggest benefit of these power modules is that they create power with next to no contamination, as the hydrogen and oxygen used to produce the power combine to deliver water as a result. Cells that utilize unadulterated hydrogen as fuel are totally carbon free.

Energy component frameworks that use hydrocarbon energizes, for example, gaseous petrol, biogas, or methanol, are among the various types. Since fuel cells utilize an electrochemical response instead of burning, they can accomplish higher efficiencies than conventional energy creation strategies. This can be worked on further with consolidated intensity and power generators that utilize squander heat from the cell for warming or cooling applications.

The interaction by which a power module works can be summed up as follows:

- Hydrogen ions enter at the anode while oxygen is taken care of at the cathode.
- The hydrogen particles are isolated into protons and electrons at the anode.
- The now decidedly charged protons go through the layer (or electrolyte) to the cathode, with the adversely charged electrons taking an alternate course as they are constrained through a circuit to create power.
- After going through the circuit and the film as needed, the electrons and protons meet at the cathode, where they join with oxygen to deliver intensity and water as side-effects.

Single energy components don't produce a lot of power, so they are organized into stacks to generate sufficient power for their expected purpose, whether that is driving a little advanced gadget or a power plant. Power modules work like batteries. Be that as it may, dissimilar to batteries, they won't run down or need re-energizing and can keep on creating power while the fuel source (for this situation, hydrogen) is provided. With the inclusion of an anode, cathode, and an electrolyte layer, there are no moving parts in a power device, making them quiet in activity and profoundly dependable.

Hydrogen energy vehicles combine the reach and refueling of traditional vehicles with the sporting and natural advantages of driving on power. Refueling an energy unit vehicle is tantamount to refueling a customary vehicle or truck; compressed hydrogen is sold at hydrogen refueling stations, taking under 10 minutes to fill current models. A few leases might take care of the expense of refueling completely. When filled, the driving scopes of an energy unit vehicle change, yet they are like the scopes of gas or diesel-just vehicles (200-300 miles). Contrasted with battery-electric vehicles, which re-energize their batteries by connecting, the blend of quick, brought-together refueling and longer driving reaches makes energy components especially suitable for bigger vehicles with significant distance prerequisites, or for drivers who need module access at home.

CHAPTER-5
WHAT IS LIQUEFIED PETROLEUM GAS AND HOW IT IS PRODUCED?

What is liquefied petroleum gas?

Liquefied petroleum gas, or LPG, is a sort of hydrocarbon gas that is acquired by refining raw petroleum or handling flammable gas. This gas is made out of one or the other, propane or butane, without help from anyone else, or a combination of the two. Aside from its use as a fuel for cooking and heating, LPG is also important for use in manufacturing applications as a fuel for vehicles, and it is frequently used to control co-generation plants.

How liquefied petroleum gas is produced?

Liquefied petroleum gas is produced during the refining process of raw petroleum or removed during the handling of flammable gas. The gases produced in this interaction are mostly propane and butane, with limited quantities of other gases. These gases are melted through compression to make them simpler to move and store.

To melt the fuel, gases are put away in durable tanks and held at high pressures—multiples of the climatic strain. These tanks have extra security highlights due to this outrageous compression, mostly an underlying shutoff valve to seal the tank assuming there are releases and an additional solid design. Since LPG by and large has no scent, limited quantities of ethanethiol or ethyl mercaptan are added to assist people in smelling hazardous gas spills.

Usage of LPG

LPG has a high caloric value, implying that it is a decent energy source as it gives an elevated degree of intensity. It is also an important fuel as it has practically no sulfur content, which means cleaner burning. A portion of the LPG utilized is consumed for warming and cooking and is basically utilized instead of petroleum gas. The leftover half of LPG is divided pretty much similarly between use in vehicles and for modern purposes.

When utilized, LPG is, for the most part, conveyed by trucks in an enormous tank and put outside of a home or other structure. Also, reusable gas canisters are accessible for controlling ovens, radiators, and grills. Little canisters of LPG are also accessible for compact hair styling tools.

Two significant drawbacks to the utilization of LPG are well-being and cost. The high strain expected to store LPG brings about periodic tank explosions in the event that canisters are not put away as expected and kept up with. Likewise, LPG is exceptionally combustible. However, providers avoid potential risk to guarantee that LPG is basically as protected as customary petroleum gas supply. The expense of LPG is a few times higher than normal gaseous petrol, yet LPG could end up being a decent choice in the event that admittance to flammable gas isn't available.

Usage of LPG as a Vehicle Fuel

LPG can be utilized as an elective fuel to control gas-powered motors as it is more neatly consumed than gas and can deliver lower measures of a few destructive emanations, for example, carbon dioxide.

One of the significant advantages of using LPG as an elective fuel is that it is typically more affordable than gasoline. Additionally, the cautious planning and well-being highlights of the tanks make them marginally more secure to use than gas, essentially on the grounds that they have a turned-down valve. This limits the chance of an LPG fire whenever it is utilized as a vehicle fuel. Like LPG, it can likewise be utilized as a fuel without detracting from vehicle performance.

Nonetheless, the accessibility of vehicles that are LPG-energized is restricted. Some existing vehicles can be switched over completely to utilize this LPG with establishments, but this requires introducing a different fuel framework as the fluid is put away in profoundly compressed fuel tanks. Moreover, it is harder to track down spots to top off LPG as it isn't quite so widely utilized as gas or diesel, and fewer miles can be driven on a solitary tank of LPG.

CHAPTER-6
HOW HYDROELECTRIC POWER PLANT WORKS?

A power plant which utilizes the energy of falling water in order to generate electricity is called hydroelectric power plant. A turbine of the plant converts the potential energy of falling water into kinetic energy. Kinetic energy is a form of energy which is produced due to the motion of an object or particle. Then a generator of the plant converts this kinetic energy from the turbine into electrical energy.

Parts of a hydroelectric power Plant

Most regular hydroelectric power plants incorporate following four significant parts:

1. **Dam:** Raises the water level of the stream to make falling water. Likewise controls the progression of water. The repository that is shaped is, as a result, put away energy.
2. **Turbine:** The energy of falling water pushing against the turbine's sharp edges makes the turbine turn. A water turbine is similar to a windmill, with the exception that the energy is produced by falling water rather than wind. The turbine converts the potential energy of falling water into kinetic energy.
3. **Generator:** Associated with the turbine by shafts and conceivably outfits so when the turbine turns it makes the generator turn moreover. It converts the kinetic energy from the turbine into electrical energy. Generators in hydroelectric

power plants work very much like the generators in different kinds of power plants.

4. **Transmission lines:** Transmission lines are used to transmit power from the hydroelectric power plant to homes and business.

How much power a hydroelectric power plant can produce?

Production of power by a hydroelectric power plant relies upon two factors:

1. **How Far the Water Falls:** The farther the water falls, the more power it has. By and large, the distance that the water falls relies upon the size of the dam. The higher the dam, the farther the water falls and the more power it has. Researchers would agree that that the force of falling water is "straightforwardly relative" to the distance it falls. As such, water falling two times as far has two times as much energy.
2. **Amount of Water Falling:** More water falling through the turbine will create more power. How much water accessible relies upon how much water streaming down the waterway. Greater streams have seriously streaming water and can create more energy. Power is too "straightforwardly relative" to stream. A stream with two times how much streaming water as one more waterway can create two times as much energy.

How to calculate the output energy of a dam?

We can compute the energy of a dam using following formulas:

Power = (Dam Height) x (Flow of Water) x (Efficiency) / 11.8

Where,

- ➤ **Power:** The electric power in kilowatts (one kilowatt rises to 1,000 watts).
- ➤ **Dam Height:** The distance the water falls estimated in feet.
- ➤ **Flow of Water:** The measure of water streaming in the waterway estimated in cubic feet each second.
- ➤ **Efficiency:** How well turbine and generator of the plant convert the energy of falling water into electrical energy. This efficiency ranges from 60% (0.60) to 90% (0.90) based on the product quality of the turbine and generator.
- ➤ **11.8:** Converts units of feet and seconds into kilowatts.

Let's consider that the dam height is 9 feet, the flow of water is 400 cubic feet each second) and we have a turbine and generator with an efficiency of 80%. In this case, the power for the dam will be:

Power = (9 feet) x (400 cubic feet each second) x (0.80)/11.8 = 244 kilowatts

To get a thought what 244 kilowatts implies, we should perceive how much electrical energy we can make in a year. As the unit of electrical energy basically in kilowatt-hours, the power from the dam is multiplied by the quantity of hours in a year.

Electrical Energy = (244 kilowatts) x (24 hours of the day) x (365 days out of every year) = 2,137,440 kilowatt hours.

CHAPTER-7
HOW ELECTRIC CARS WORK?

Cars that can run without internal combustion engines are called electric cars. These cars are equipped with a rechargeable battery pack and an electric motor. There are no harmful exhaust emissions or burning up of gasoline during the driving of these cars. Noise pollution is also very low in these cars. These cars are energized by plugging into a charging point at an electric charging station. Energy for the car is stored in a rechargeable battery, which provides necessary power to the motor for turning the wheels of the car. These cars move comparatively faster than traditional fuel based cars.

The Core Components of an Electric Car

An electric car contains the following core components:

Charging Port

The charging port of an electric car works as a medium to supply energy to the battery from an electric charging station. A standard 240-volt outlet can charge an electric car for the whole night.

Inverter

In order to convert the direct current (DC) of the battery of an electric car into an alternating current (AC), inverters are used. After converting into AC, the necessary power is supplied to the electric traction motor. The motor frequencies are controlled by the inverter. Therefore, there is a significant impact of the inverter on controlling the speed of the electric car.

Electric Motor

A rotating magnetic field is generated when the electric traction motor gets AC power from the inverter. It helps turn the motor. The efficiency of the motor is comparatively high. It provides power from the pedal to the engine instantly.

Battery

The supplied energy of the car is stored in the battery, which helps to provide necessary power to the motor and other electrical components. Most cars use lithium-ion batteries, which provide a huge amount of current and require less maintenance in comparison to other batteries. The unit of power is kilowatts (kW). On the other hand, the unit of energy is the kilowatt-hour (kWh), which indicates how much electric power is used for an hour. An electric car consumes an average of around 2,000 kWh of energy a year.

The electric power-train

An electric power-train basically consists of an inverter, an electric motor, a reduction drive, and a battery. It incorporates the overall high-voltage electrical system for driving the car. They are compressed and deliver less vibration with instant torque. Some inverters in the power-train help to transfer unused AC power during braking into DC power. This power can be stored back in the battery.

Types of electric cars:

There are various types of electric cars. Cars that run only on electricity are known as pure electric cars. On the other hand, cars that can also operate on diesel or petrol are called hybrid electric cars.

- **Plug-in electric cars:** These cars operate only on electric power. They are energized when they are plugged into a charging point. As there is no requirement for diesel or petrol to operate, there are no emissions in these cars.
- **Plug-in hybrid cars:** These cars basically operate on electric power but also have a conventional fuel engine wherein fuel such as diesel or petrol can be used in case of a shortage of charge. Emissions are produced during fuel operation. This type of car can be plugged into an electric power source for recharging the battery.
- **Hybrid electric cars:** These cars operate essentially on fuel such as diesel or petroleum or diesel and also have the option of an electric battery. There are options with a button for switching from fuel engine mode to electric car mode. These cars depend on fuel, such as diesel or petrol, for energy. There is no option to plug into an electric power source.

How to charge an electric car?

An electric car can be charged either by connecting it to an attachment or by connecting it to a charging unit. There are three sorts of chargers:

- Three-pin plug: a standard three-pin plug that can be interfaced with any 13-amp attachment.
- Socketed: a charge point which can be interfaced with either a Type 1 or Type 2 link.
- Fastened: an accuse point of a link appended with either a Type 1 or Type 2 connector.

Charging speed

There are, additionally, three types of charging speeds:

- Slow: Has a charging time of 8–10 hours and can charge up to 3kW.
- Quick: Can be charged either at 7Kw or 22kW with a charging time of 3–4 hours.
- Charges very quickly: Can charge up to 43 kW in 30 to 60 minutes. It is only compatible with electric cars that have very fast charging ability.

Charging up in various seasons

The weather conditions influences how much energy your electric can consumes. You have a bigger reach in summer and more modest reach in winter.

CHAPTER-8
WHAT IS MICROCONTROLLER AND HOW IT WORKS?

What is microcontroller?

Microcontroller is a close-packed integrated circuit which can control a specific operation in an embedded system. It includes a memory, processor and input/output (I/O) peripherals on a single chip. As Microcontrollers can control the actions and features of a product, they are also known as embedded controllers. They can run one specific program and operate a single task. Microcontroller consists of input and output device. It has small LED or LCD display outputs. The power consumption of microcontroller is very little. Microcontrollers take inputs from the device and keep controlling the device by sending signals to different parts of the device. For example, we can say the microcontroller of a television. It takes input from a remote control and delivers output on the television screen. Microcontrollers are also found in robots, office machines, vehicles, medical devices, mobile radio transceivers, vending machines, home appliances and other devices. Microcontroller is basically a simple miniature personal computers (PCs) which can control small characteristics of a larger component without a complex front-end operating system (OS).

Components of a microcontroller

A microcontroller consists of different components in order to perform its tasks smoothly. Some of these components are as follows:

RAM

Full abbreviation of RAM is Random Access Memory. Data and output are stored in RAM. These data are stored temporarily. Whenever, the power supply to the microcontroller, the memory of RAM is lost. There is a special function register (SFR) in RAM. There is a pre-configured memory called special functions register (SFR) in RAM which is offered by the manufacturer of the microcontroller. Behavior of the serial communication and analog-to-digital converter are controlled by it.

ROM

Full abbreviation of ROM is Read Only Memory. It stores special tasks of microcontrollers which can never be changed. ROM helps microcontrollers to understand that certain actions should trigger particular responses. For example, ROM helps the microcontroller of television to understand that pressing a channel button should change the display on your screen. The program size which is stored on ROM depends on the ROM's size. Some microcontrollers accept ROM addition in the form of external chips while others come with built-in ROM.

Program counter

Depending on a series of different programmed instructions, a program counter assists the minicomputer for executing programs The program counter can keep tracking the counter's place in the line of code by increasing the executed instruction one every time.

Inputs and outputs

Microcontrollers can interact with humans through inputs and outputs in a special way. LED displays, temperature, humidity and light levels indicating switches and sensors are some of input and output devices on a microcontroller. A wide range of input and output pins or GPIO are configured for different input and output devices on a microcontroller. For example, one pin configured as an input on the microcontroller works by sensing temperature and another pin configured as the output and connected to the thermostat triggers the air conditioner or heater to turn on and off based on the pre-set temperature ranges. Input and output dynamics are completely machine-to-machine. There is no need of direct human interactions for taking any decision.

How does a microcontroller work?

A microcontroller receives data from its input/output (I/O) peripherals using its central processor and interprets the data. In this way. it controls a singular function in a device. Temporary data or information are stored in its data memory. In this scenario, the processor accesses it and uses instructions stored in its program memory for decoding and applying the incoming data. After then, it communicates and validates the appropriate action by using its I/O peripherals. Multiple microcontrollers can work together handling their respective tasks within a device. For example, we can say about a car. A car consists of many microcontrollers which can control various individual systems within, like the traction control, anti-lock braking system, fuel injection or suspension control by communicating with each other to inform the correct actions. Some are capable to communicate with a more complex central computer within the car and others can only communicate with other microcontrollers. In this case, they send and receive data using their I/O peripherals help to send, receive and process the data for performing their assigned functions.

CHAPTER-9
WHAT IS MICROPROCESSOR AND HOW IT WORKS?

A microprocessor is a programmable electronics chip which can compute and make decision as like as the central processing unit (CPU) of a computer. It can process and execute the instruction using its control unit of arithmetic and logic unit (ALU). Microprocessors are used in mobile phones, washing machines, printers etc. It can perform various arithmetic operations like sum, subtraction, division, multiplication, and more. It can also perform logical operations such as AND, OR, greater than, less than, and more.

A typical microprocessor consists of arithmetic and logic unit (ALU) in association with control unit to process the instruction execution. In order to perform any task by the microprocessor, it must be preprogrammed by the user or programmer. In this case, a programmer knows its internal resources, features and supported instructions. The microprocessor manufacturer provides a list of instructions of each processor. There are two forms of instruction set in a microprocessor. They are binary machine code and mnemonics. Microprocessor runs and communicates with binary numbers of 1 and 0. Binary pattern instruction is called a machine language which is difficult for us to understand. So, the binary patterns are given abbreviated names, called mnemonics, which forms the assembly language. For converting assembly-level language into binary machine-level language, is an application is used which is known as assembler.

The first successful processor used in practical applications is the microprocessor 8085 which is designed by Intel in 1977 and manufactured by using NMOS Technology. The 8085 microprocessor is used in different devices and projects like automation systems, digital controllers, calculators, video game players, embedded systems, smartwatches, etc.It is an 8-bit general purpose microprocessor which can operate at a +5V DC power supply with a 3.2 MHz single-phase clock. The microprocessor 8085 works based on the following components and internal parts:

1. Arithmetic and Logic Unit (ALU)

ALU is the core part of the microprocessor which can perform arithmetic or mathematical operations like addition, subtraction, multiplication, division, and logical operations like AND, OR, NOT, increment and decrement. It works with 8-bit data only. It takes data from the accumulator and temporary register during executing the program.

2. Accumulator

In order to perform the arithmetic and logical operations, an 8-bit register supplies the data directly to the ALU during executing the program. It is called accumulator which is connected between the internal bus and ALU. Accumulator can load and store new data.

3. General Purpose Registers

There are six general-purpose registers named as B, C, D, E, H, and L in the microprocessor 8085. Each of them can store 8-bit data. Pairs like B-C or D-E, or H-L can store 16-bit data.

4. Flag Register

Values either 1 or 0 are stored in this register. It's also an 8-bit register which can only store these values based on the value is stored in the accumulator after executing the program.

5. Temporary Register

This is an 8-bit register wherein data are temporarily stored during executing the program. It provides supporting data to the ALU. In this case, the data comes from the general-purpose register and is stored in the temporary register during execution of the program by ALU.

6. Instruction Register and Decoder

Instruction register stores instruction fetched from the memory during the program execution. The instruction decoder decodes the instruction to understand the task to be performed by the ALU.

7. **Program Counter**

It is a 16-bit register which stores the memory address of the instruction that will be executed in the next. The program counter will be increment by one whenever the instruction is executed.

8. **Stack Pointer**

A 16-bit register which works as a stack and always makes increments or decrements by 2 during the push and pop operation is called stack pointer. The content of the register is stored here for using in the next execution.

9. Timing and Control Unit

In order to perform tasks or to do program executions, timing and control unit provides all the clock or pulse signals to all the components of the microprocessor. It is the combination of analog and digital circuits. It not only provides signals to the internal components but also provides the timing and control signal to the external component as well as circuit connected to the microprocessor like RD, WR, HLDA, HOLD, READY, etc.

10. Interrupt Control

The control system which controls the interrupts in the microprocessor during executing the program is called Interrupt Control. ALU stops the current program execution and processes the task given by the interrupt if there is an interrupt occurs by an external or internal input. It starts again its own work after finishing the task given by the interrupt.

11. Serial Input/output Control

A system which can control the serial communication between the microprocessor and external devices through serial input and serial output ports like SOD and SID is called Serial Input/output Control.

12. Data Bus and Address Bus

For transferring data between memory and processor or between I/O device and processor, a bus is used which is called data bus. It is a bidirectional bus which can carry and transfer the data is to be stored. 8-bit processor contains an 8-bit data bus and a 16-bit processor has 16-bit data bus. There are 8-bit data bus and 16-bit address bus in the 8085 microprocessor. On the other hand, the address bus carries the location address of the memory where the data is to be stored. It is a unidirectional bus. In this case, a unique address or binary pattern is used for identifying a memory location or an input/output (I/O) port. For example, an 8 bit address bus contains eight lines having address 2^8 = 256 different locations. The locations in hexadecimal format can be written as 00H – FFH.

CHAPTER-10
WHAT IS BIOMASS AND HOW TO CONVERT BIOMASS INTO ELECTRICITY?

What is Biomass?

Biomass is the biological component obtained from living organisms. It is also referred as plant based material when energy is produced from biomass. At the same time, biomass is also applicable for both animal and vegetable resultant components.

Biomass is a renewable energy source and it can re-grow quite quickly. Sun's energy is captured by plants' chlorophyll through the photosynthesis process. Here, carbohydrates are created by converting carbon dioxide from the air and water from the ground. They are complex combination of carbon, hydrogen and oxygen.

Basically, plant materials are either broken down by micro-organisms or burned. After burning, carbohydrates are transformed into carbon dioxide and water and release the energy harnessed from the sun. This process is known as the carbon cycle in the earth. In this technique, biomass works like natural battery in order to store solar energy.

Appropriate utilization of biomass can enhance the longevity of this natural battery as well as provide low-carbon energy sources. Sustainable low-carbon biomass is great source of new renewable energy which is environmentally friendly indeed.

Types of beneficial biomass

There are following types of beneficial biomass in this planet:

- Wood and forest residues which are harvested in sustainable way.
- Crop residues like wheat straw.
- Energy crops which don't compete with food crops for land.
- Clean industrial and municipal wastes.

Typically, beneficial biomass sources help to maintain and increase the stocks of carbon stored in plants or soil. It also moves emissions from fossil fuels, like coal, natural gas and oil. Most sustainable and effective biomass resources may be varied from region to region. It also depends on the types of final application you wish to use. For example, you can utilize biomass as bio power, bio products, heat and biofuel.

Benefits of biomass

- Typically, biomass offers environmental, economic and energy security benefits.
- Biomass energy decreases air pollution and net carbon emissions.
- It contributes to protect the quality of soil, maintain wildlife habitat and avoid erosion.

- Biomass helps to minimize the dependency on importing fossil fuels which reduces our expenses.
- Rapid growth of biomass energy technology helps farmers and forest owners to obtain valuable new markets for their new energy crops, forest and crops residues.
- It contributes to minimize the global warming.

How to convert biomass into electricity?

Generating heat energy by burning biomass is one of the most familiar method since long time. This biomass fired heat can produce steam power for generating electricity. Burning biomass in conventional boilers provides environmental and air-quality benefits over burning fossil fuels.

Recent researches by researchers indicate that biomass can be converted more cleanly into liquid fuels. For example, producing combustible gases from gaseous process. It decreases different kinds of emissions from biomass combustion.

Technologies which can convert renewable biomass fuels into heat and electricity is called biopower technologies. It follows same process wherein fossil fuels are used to generate electricity. The energy is stored in biomass. It can be released to produce biopower by using three methods. They are – Direct Combustion, Decomposing by Anaerobic Bacteria and conversion to gas/liquid fuel.

Direct combustion

Direct combustion is one of the most used methods for generating electricity from biomass. In this method, high-pressure steam is produced by burning biomass. This steam flows over a series of turbine blades and help them to rotate. The rotation of the turbine drives a generator for generating electricity. In an existing power plant furnace wherein a portion of coals are used for generating electricity, there biomass can be used as substitute by co-firing process. Co-firing is combusting two different types of materials at the same time.

Decomposing by anaerobic bacteria

Organic waste material animal dung or human sewage, is collected in oxygen-free tanks called digesters. Some organic waste materials like cow dung, human sewage etc. are reserved in oxygen-free tanks called digesters and decomposed by anaerobic bacteria. It produces methane and other byproducts to form a renewable natural gas. After purifying, it is used for generating electricity.

Converting biomass into to a gas or liquid fuel

By using gasification and pyrolysis process, biomass can be converted to a gaseous or liquid fuel. In the gasification process, synthesis gas or syngas is produced by exposing solid biomass material to high temperatures with very small amount of oxygen present. Synthesis gas or syngas is a mixture that consists mostly of carbon monoxide and hydrogen. This gas is burnt in a conventional boiler for generating electricity. In a combined-cycle gas turbine, it can also be used as a replacement of natural gas.

In the process of pyrolysis, crude bio-oil can be produced by heating biomass in the complete absence of oxygen at a lower temperature range. This bio-oil can be used in the power plant as a substitute wherein diesel or fuel oil is used for generating electricity.

CHAPTER-11
HOW SOLAR CELLS WORK?

Solar cell is an electrical device which converts the light energy directly into electricity utilizing photovoltaic effect. It is also defined as the form of photoelectric cell having electrical characteristics like current, voltage and resistance.

There are at least two semiconductor layers in solar cells. One layer contains a positive charge and the other layer contains a negative charge. A typical silicon solar cell consists of a thin wafer having phosphorus doped (N-type) ultra thin silicon layer on the top of boron-doped (P-type) thicker silicon layer. When these two materials are connected with each other, a junction is formed which is called P-N junction. As a result, an electrical field is created near the top surface of the cell. When these two layers are connected to an external load, the electrons flow through the circuit generates electricity.

Sunlight consists of small particles of solar energy called photons. When sunlight strikes on the surface of solar cell, many of the photons are reflected and absorbed by the solar cell. Electrons are released from the negative layer of semiconductor material, when enough photons are absorbed by this layer of the solar cell. These electrons naturally move into the positive layer and create a voltage differential.

Under open circuit, no load conditions, a typical silicon solar cell generates about 0.5 – 0.6 volt DC (Direct Current). The output of a solar cell depends on its surface area (size) and efficiency. It is also proportional to the light intensity of the sun striking the surface of the cell. For example, under peak sunlight conditions, a typical commercial solar cell with a surface area of 160 square centimeters will approximately generate peak power of 2 watts. If the intensity of the sunlight is 40 percent of peak, then the cell would generate approximately 0.8 watts. In order to increase the output power, cells are combined in a weather tight package which is called a solar module. In order to create the desired voltage and amperage, these modules (from one to several thousand) are then wired up in series and parallel with each other. It is called a solar array.

The semiconductor material silicon which is primarily used for the manufacturing process of solar cells is naturally available. Due to the natural availability of silicon and the practically unlimited resource in the sun, solar cells are very environmentally friendly. Solar cells burn no fuel and have absolutely no moving parts which makes them virtually maintenance free, silent and clean.

CHAPTER-12
HOW SOLAR PANELS ARE MANUFACTURED?

A solar PV module contains solar cells, glass, EVA, back sheet and frame. Three types of solar panels are available in the market. They are:

- ➢ Mono-crystalline solar panels
- ➢ Poly-crystalline solar panels
- ➢ Thin film solar panels

Various types of materials are used for manufacturing at cell structure level. They are - mono silicon, poly silicon or amorphous silicon. Mono and Poly crystalline cells have almost similar manufacturing process. For manufacturing a crystalline solar panel following steps are followed:

First Step: Sand

Here sands are used as a raw material. Most solar panels are made of silicon, which is the main component in natural beach sand. Silicon is plentifully available which is the second most available element on Earth. However, converting sand into high grade silicon is a high cost energy intensive process. Pure silicon is produced from quartz sand in an arc furnace at very high temperatures.

Second Step: Ingots

Basically, the silicon is collected in the form of solid rocks. Hundreds of these rocks are being melted together at very high temperatures in order to form ingots in the shape of a cylinder. A steel, cylindrical furnace is used for forming desired shape. All atoms need to be perfectly aligned in the desired structure and orientation during melting process. For providing the silicone positive electrical polarity, Boron is added to the process

Mono crystalline cells are manufactured from a single crystal of silicon. Mono Silicon has higher efficiency in converting solar energy into electricity, therefore the price of mono crystalline panels is comparatively higher.

Poly silicon cells are made from melting several silicon crystals together. After the ingot has cooled down, grinding and polishing are being performed, leaving the ingot with flat sides.

Third Step: Wafers

In this step, wafers are used during manufacturing process. The silicon ingot is sliced into thin disks, also called wafers. A wire saw is used for precision cutting. The thinness of the wafer is similar to that of a piece of paper. As pure silicon is shiny, it can reflect the sunlight. An anti-reflective coating is put on the silicon wafer for reducing the amount of sunlight lost.

Fourth Step: Solar cells

Treating each wafer, metal conductors are added on each surface. The conductors provide the wafer a grid-like matrix on the surface. This confirms the conversion of solar energy into electricity. The coating will ease the absorption of sunlight, rather than reflecting it. In an oven-like chamber, phosphorous is being diffused in a thin layer over the surface of the wafers. This will charge the surface with a negative electrical orientation. Boron and phosphorous combination will provide the positive - negative junction, which is very important for the proper function of the PV cell.

Fifth Step: From Solar Cell to Solar Panel

In this step, using metal connectors, the solar cells are soldered together to connect the cells. Solar panels are made of solar cells integrated together in a matrix-like structure. The current standard offering in the market are:

    48 cell panels - For small residential roofs.

    60-cell panels - The standard size.

    72-cell panels - For large-scale solar power plant.

After putting the cells together, a thin layer (about 6-7 mm) of glass is added on the front side, facing the sun. Highly durable, polymer-based material is used to make the back sheet. This will protect solar panel entering water, soil and other materials from the back. For enabling connections inside the module, the junction box is added.

After assembling the frame, it all comes together. The frame protects the panel from impact and weather. The use of a frame will also allow the mounting of the panel in a variety of ways, for example with mounting clamps. EVA (ethylene vinyl acetate) is the glue which binds everything together. It is crucial that the quality of the encapsulation is high so it doesn't damage the cells under harsh weather conditions.

Sixth Step: Testing the Modules

For ensuring expected performance of the cells, testing is carried in this step. perform as expected. STC (Standard Test Conditions) are used as a reference point. The panel is put in a flash tester at the manufacturing facility. The tester will deliver the equivalent of 1000W/m2 irradiance, 25°C cell temperature and an air mass of 1.5g. Electrical parameters are written down and these results can be found on the technical specification sheet of every panel. The ratings will reveal the power output, efficiency, voltage, current, impact and temperature tolerance.

Besides STC, every manufacturer uses NOCT (nominal operating cell temperature). The parameters used are more close to 'real life' scenario: open-circuit module operation temperature at 800W/m2 irradiance, 20°C ambient temperature, 1m/s wind speed. The ratings at NOCT can be found on the technical specification sheet.

Before shipping the module to homes or businesses, cleaning and inspection are done which is the final step of the production.

The aim of the research and development in the solar energy industry is to reduce the cost of solar panels and increase the efficiency. The solar panel manufacturing industry is becoming more viable and is predicted to become more popular than conventional sources of energy like fossil fuels.

CHAPTER-13
HOW DIODES WORK IN A SOLAR PANEL?

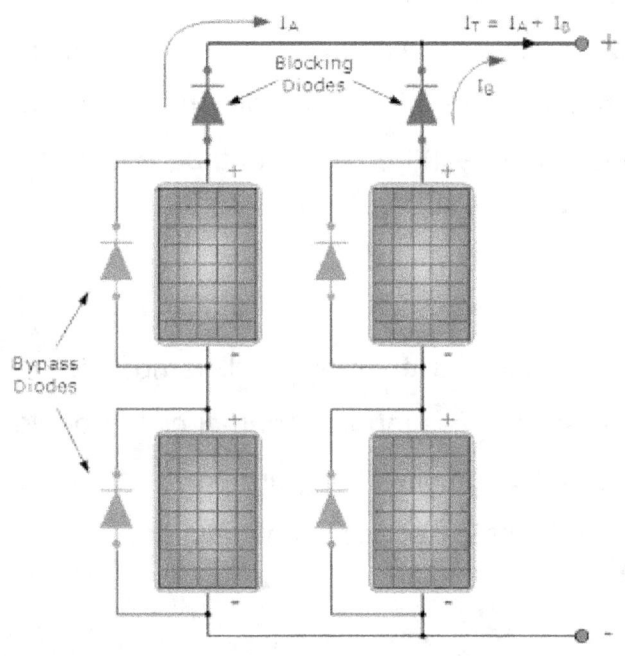

Diodes are two terminal electronic components which allow current to flow in one direction. This diode can be used to block the flow of electric current from other parts of an electrical circuit. These types of silicon diodes are basically known as Blocking Diodes during the usage with a solar panel. Bypass Diodes are connected in parallel with either a single or a number of solar cells to prevent the solar cells overheating by sunlight. By providing a current path around the bad cell, these types of diodes also protect the partially shaded solar cells from burning out. Blocking diodes are used differently than bypass diodes.

Bypass diodes in solar panels are connected in "parallel" with a solar cell or panel to shunt the current around it, whereas blocking diodes are connected in "series" with the PV panels to prevent current flowing back into them. Therefore, blocking diodes are different than bypass diodes. These types of diodes are physically the same in most cases. In order to serve various purposes, they are used in different way.

Usage of Bypass Diodes in Solar Photovoltaic Arrays

Here, diodes with green color are "bypass diodes", one in parallel with each solar panel to provide a low resistance path. Bypass diodes in solar panels and arrays safely carry this short circuit current. On the other hand, diodes with red colors are known as the "blocking diodes", one in series with each series branch. These blocking diodes are also known as a series diode or isolation diode, ensure that the electrical current only flows in one direction "OUT" of the series array to the external load, controller or batteries.

The reason for this is to prevent the current generated by the other parallel connected solar panels in the same array flowing back through a shaded solar cell and also to prevent the fully charged batteries from discharging or draining back through the array at night. Therefore, when several solar panels are connected in parallel, blocking diodes are used in each parallel connected branch.

Blocking diodes are basically used in solar photovoltaic arrays when there are two or more parallel branches or there is a possibility that some of the array will become partially shaded during the day as the sun moves across the sky. The size and type of blocking diode used depends upon the type of solar photovoltaic array.

Two types of diodes are available as bypass diodes in solar panels and arrays. One is the PN-junction silicon diode and another is the Schottky barrier diode. Both diodes are available with a wide range of current ratings. Forward voltage drop of the Schottky barrier diode is about 0.4 volts/ On the other hand, the PN junction silicon diodes have the voltage drop of 0.7 volt for a silicon device.

This lower voltage drop helps to save one full solar cell in each series branch of the solar photovoltaic array. Because of dissipating less power in the blocking diode, the array becomes more efficient. During manufacturing solar panel, most manufacturers add both bypass and blocking diodes in their solar panels.

CHAPTER-14
WHAT IS OCEAN ENERGY AND HOW DOES IT WORK?

What is ocean energy?

All type of renewable energy which is acquired from the sea is called ocean energy. Constant flow of ocean currents contains huge amount of water across the earth's ocean. Technological development contributes to extract energy from ocean currents and convert it into usable power.

Constantly moving ocean waters are affected by water salinity, wind, rotation of the earth, temperature and topography of the ocean floor. Wind and solar heating of surface water near the equator contribute to drive most ocean currents. Meanwhile, salinity and density variations of water column create some currents. Ocean currents are comparatively constant and flow in one direction. Due to the density of water, ocean currents move slowly and contain great amount of energy in comparison to typical wind speed. Water is denser than air. For this characteristics, ocean energy can be captured and converted into usable form of electricity.

Ocean Energy is classified in three types. They are as follows:

- Mechanical energy from the waves that is wave energy.
- Mechanical energy from the tides that is tidal energy.
- Thermal energy from the sun's heat.

How does it work?

1. **Wave energy:** This type of energy is produced by converting the energy of ocean waves or swells into other type of energy which is only electricity at present. Various types of technologies are developing and trialing in order to convert the energy of waves into electricity.

2. **Tidal Energy:** Tidal movements contribute to generate tidal energy. Both potential energy and kinetic energy exist in tides. Potential energy is related to the vertical oscillations in the sea level. On the other hand, kinetic energy is related to the horizontal motion of the water. Technologies of producing energy from the rise and fall of the tides can be used to harness it.

3. **Ocean Thermal Energy:** By converting the temperature difference between surface water and water at depth into useful energy, ocean thermal energy can be generated. Ocean thermal energy conversion (OTEC) is a system of twenty-four-hour base load electricity generation per day throughout the year. In

comparison to other ocean energy sources, OTEC is one of the continuously available renewable energy resources which contribute to base load power supply.

CHAPTER-15
WHAT IS GEOTHERMAL ENERGY AND HOW IS IT USED TO GENERATE ELECTRICITY?

Geothermal energy is the process of gaining heat as energy source from earth's surface. Most of power plants require steam for generating electricity. Geothermal power plants use steam which is produced from hot water tanks discovered far below earth's surface. A turbine is rotated by steam which helps to run a generator for electricity production. Most power plants are using conventional fuels to boil water for steam. Earth's surface holds huge amount of heat which can produce extremely huge amount of energy than all other natural resources like oil and gas in the world. Predicted heat extraction in geothermal energy is very small in comparison to heat content of earth. So, it is renewable.

Types of the geothermal power plant and how they work:

Geothermal power plants are of three types. They are - flash steam, binary cycle and small scale power plant.

Flash steam power plants: Here, geothermal tanks of water having temperatures higher than 183°C are used. This hot water flows up with its self-pressure through shafts in the ground. Pressure decreases with its upward flowing and some portion of hot water boils into steam. The steam is then separated from the water and used to rotate the turbine in order to generate power. Any excess water and compressed steam are injected back into the tank which is a significant example of sustainable resource.

Binary cycle power plants: These plants work on water at lower temperatures of about 108°-182°C. In order to boil a working fluid, these plants utilize the heat of hot water which is basically an organic compound having lower boiling point. In the heat exchanger, water fluid is evaporated which helps to rotate a turbine. For reheating, the water is then injected back into the ground. During whole process, working fluid and the water are kept separated. So, there is no possibility of air emissions.

Small-scale power plants: These plants are typically under 5 megawatts. They have good prospect of extensive application in rural areas as well as a distributed energy resource. Distributed energy resource is the diversity of small and flexible power producing technologies which is integrated to develop the process of the electricity delivery system.

Applications of Geothermal Energy:

Long time ago, geothermal energy was used for bathing and space heating. Now, it is known for generating electricity. It can also be used for industrial processes, purification and pumping system.

Benefits of Geothermal Energy:

No fuel is required in geothermal energy and its capital costs are very significant. So, it is very cost effective. It is also environmentally friendly, consistent and sustainable. Historically, geothermal energy is limited to areas close to boundaries of tectonic plate. Many areas of the world are already using geothermal energy as a reasonable and ecological solution. At present, countries like United States, Japan, New Zealand, Italy, Iceland, Mexico, El Salvador, Philippines, Indonesia and Kenya are generating the most electricity from geothermal sources. This technology helps to minimize the dependency on fossil fuels and also reduces the global warming.

CHAPTER-16
WHAT IS WIND ENERGY, HOW DOES WIND POWER WORK AND HOW TO DESIGN A WIND FARM?

What is wind?

Wind is the movement or flow of gaseous elements around earth surface. One of the most important characteristics of wind is its velocity. Another characteristic of wind is energy. It is affected by high and low pressure of air of that particular location or area. Earth surface is heated randomly by the sun which depends on the occurrence of sun ray's angle and surrounding of the land. Sun ray's angle can be varied with latitude and time of day. Water of the oceans heat up and cool down slowly in comparison to land. Heat energy which is absorbed in earth's surface is transferred to air directly above these oceans and lands. The density of warmer air is less in comparison to cooler air. These factors are creating constantly changing characteristics of wind across earth's surface.

What is wind energy?

Energy obtained from the wind force is called wind energy. It transforms the kinetic energy of air flow into electrical energy by using wind turbines. The energy is mainly extracted with the rotor which transforms the kinetic energy into mechanical energy and it transforms this mechanical energy into electrical energy with the generator. Wind energy is a renewable energy which is environment friendly. Economically generating wind power is the total amount of economically extractable power available from the wind. It is significantly higher than power generated from all other sources.

How does wind power work?

It is necessary to use a wind turbine for utilizing the kinetic energy of the wind and convert it into electrical energy, Wind turbines are basically 80 to 120-meter-high based on the strength of the wind. In order to obtain this energy in large quantities, wind turbines must be set up in places where windy conditions are predominant.

The wind turbines have to be oriented in the direction of the wind, which is done by means of a vane on the nacelle. From there, the force of the air currents will set the three main parts of the wind turbine in motion:

> **The rotor:** It is composed of three blades and the bushing which joins them together. It captures the force of the wind and convert it into mechanical rotational energy.
> **The multiplier:** It is connected to the engine by means of a shaft. It increases the rotational speed from 30 revolutions per minute (rpm) to 1500 rpm.
> **The generator:** It converts the mechanical energy of rotation into electrical energy.

Wind turbines of a wind farm are connected together by underground cables which carry the electricity to a transformer substation.

How to design a wind farm?

A Wind Farm or Wind Park is the assembly of wind turbines which helps to generate power from wind in the same area. Typically, a large wind farm has some individual wind turbines which covers prolonged areas. But, the land between the turbines should be utilized for agricultural or other purposes. It may also be constructed in offshore area.

There should be a constant flow of non-choppy wind throughout the year having less possibility of stormy wind in the location of a wind farm. Basically, economic wind generators need wind speed of approximately 16 km/h (10 mph) or greater. An important factor of turbine siting is also access to local demand or transmission capacity.

Basically, a wind farm is planned and designed on the basis of a meteorological wind data charts as well as practically measuring actual wind speed of the area. In order to finance the project, it is very important to survey the area and collect the data of wind speed of that particular area. Sometimes, wind in local areas observed for a year or more. Before installing wind generators and detailed wind maps should be prepared.

Each turbine is connected to medium voltage through a combined power system and communications network. In order to increase the voltage of the power system, a step up transformer is used in a substation. Installation of collector system and substation are required to build up a land-based wind farm. A road access for each turbine is also required in the farm area.

CHAPTER-17
HOW DOES ARTIFICIAL INTELLIGENCE WORK?

Artificial Intelligence (AI) system is a process of reverse-engineering human behaviors and competences in a machine. It has computational ability. By using computational ability, it can exceed what human beings are capable of. Combining huge amounts of data with fast, iterative processing and intelligent algorithms, allowing the software to learn automatically from design and configurations in the data are the process to develop Artificial Intelligence. It works in following sub fields:

- In machine learning, it demonstrates a machine how to implicate and take decisions based on past experience. It classifies designs, analyses past data to conclude the meaning of these data points to reach a possible conclusion without involving human experience. This automation system helps in business by saving human time and assist human to make appropriate decision.
- In deep learning, it explains a machine to process inputs through layers in order to categorize, conclude and forecast the outcome.
- In neural networks, it processes data like human brains or neural cells. These are a series of algorithms which develops the relationship between various fundamental variables.
- In natural language processing, it helps a machine to understand what the user expects to communicate and it responds accordingly.
- In computer vision, it helps the machine to classify and learn from a set of images, to make a better output decision based on previous observations. Here, computer vision algorithms break down an image and read different parts of the objects. In this way, it tries to understand the image.
- In intellectual computing algorithms it simulates a human brain by analyzing text, speech, images and objects in a method that a human does and tries to give the desired output.

Artificial Intelligence can be made over a different set of components and will function as a consolidation of following fields:

- Philosophy
- Mathematics
- Economics
- Neuroscience
- Psychology

- Computer Engineering
- Control Theory and Cybernetics
- Linguistics

Philosophy

Philosophy assists machines to think and understand about the nature of knowledge itself. It helps to connect knowledge and action through goal-based analysis to achieve required outcomes.

Mathematics

Artificial Intelligence algorithms help to make accurate predictions of future outcomes for taking right decision. Here, the mathematical application, probability is used.

Economics

Economics explains how people make choices according to their desired results. It contains not only money, but also many important ideas like design theory, operations research and decision processes. By using mathematics, it demonstrates how these decisions are being made at large scales along with their collective results are. Intelligent Systems are developed by these types of decision-theoretic techniques.

Neuroscience

Neuroscience demonstrates how the human brain works. Artificial Intelligence tries to duplicate the same. Here, an obvious similarity is observed. Computers are millions of times faster than the human brain, but the human brain still has the advantage in terms of storage capability and interconnections. With advancement of computer hardware and more sophisticated software, it tries to achieve the intelligence level of human brain.

Psychology

Intellectual psychology views the brain as an information processing device and works based on principles and goals. This is as like as our own developed intelligence machine. Building algorithms by code helps to power the chat bots.

Computer Engineering

Computer engineering translates all concepts and theories into a machine readable language, computes it and give result which we can understand. Artificial Intelligence systems are developed with rapid advancement of the field of computer engineering. These are based on advanced operating systems, programming languages, information management systems, tools, and state-of-the-art hardware.

Control Theory and Cybernetics

For making a system intelligent appropriately, a system should be capable to control and modify its actions to produce the desired result. It is defined as an objective function. The system moves forward based on this. It can frequently modify its actions in various environment by using mathematical computations and logic to measure and improve its behaviors.

Linguistics

The formation of natural language processing is explained in Linguistics. It helps machines to understand our syntactic language, and give result so that everyone can understand it. For understanding a language, it is required to learn the structure of sentences and have a knowledge of the subject matter and circumstance.

CHAPTER-18
WHAT IS BIOELECTRICITY IN SCIENCE?

Bioelectricity is the process of producing electromagnetic energy by living organisms. The bioelectric activity which happens throughout the human body is very necessary to life. Living cells can produce electric, magnetic and electromagnetic fields which enable the action of muscles and the transmission of information in the nerves. This is the concept of quick signaling in nerves. It produces physical processes in muscles or glands. There is some similarity among the muscles, nerves and glands of all organisms. The early development of fairly efficient electrochemical systems is the reason behind it. Scientists are concentrating on the muscles or nerves tissues like the brain, heart, eye, ear, stomach and certain glands, electric organs in some fish and potentials associated with damaged tissues.

Electric movement in living tissue is a cellular concept which depends on the cell membrane. This membrane works as a capacitor wherein energy is stored as electrically charged ions on reverse sides of membranes. The stored energy is available for quick operation. It steadies the membrane system so that it is not activated by small disturbances. Cells capable of electric movement show a resting potential wherein their interiors are negative by about 0.1 volt or less compared with the outside of the cell.

Bioelectric signals are triggered by electrically active tissues like the brain, heart or the muscles. These active tissues can cause some concentration differences in the extra-cellular fluid which includes sodium, potassium and chloride ions. This is why one can measure signals like ECG or EEG from outside the body on the surface of the skin, with the help of electrodes. An interface between the extra cellular fluid and the metal of the wire is constructed by the electrode. The electrode is a sensor which consists of a metal and often a salt-bridge. Here, the local differences of the concentration of charged ions are converted into an electrical signal. The bioelectric signal which is measured from the skin's surface is within the approximate range of 0-2000 µV (2 mV).

More electrical phenomena exist inside our body or on the electrode. Two of them are the DC (Direct Current) offset of the electrode and the 50 or 60 Hz (Hertz) mains interference or main potential. Furthermore, any measurement will display the noise which is produced by the body, the electrode impedance or the amplifier itself. Measuring potential differences between two points on the body can provide very important information regarding the electrical activity happens inside our body. All these noise and signals are dealt by the measurement configuration in such a way that the bioelectrical signals which are measured on the skin's surface are reflected in the output signal positively and cleanly.

CHAPTER-19
WHAT IS ELECTRICITY IN SCIENCE?

Our universe consists of either energy or matter. Ability to do work is called energy which comes in the form of kinetic energy or energy of movement, light, heat, solar energy, geothermal energy, hydro power, wind and electricity. To understand the electricity, we have to understand a matter called atom. Atom consists of three particles called protons, electrons, and neutrons. Protons and neutrons are in the center core or nucleus of the atom while electrons spin in energy levels called orbits or shells around the nucleus. As electrons are not firmly held to the nucleus, they have the capability to leave atoms. When electrons leave atoms, matter becomes charged. During the flow of electronics, electricity is generated. There are two types of electricity: static electricity and current electricity. Static electricity study is also known as electrostatics. When objects become charged, static electricity is generated. The charges on these objects do not flow from one object to the other. Objects become charged when they are rubbed together. When two objects are rubbed together, electrons can jump from one object to the other. Current electricity is the type of electricity which occurs when electricity flows through a specific path or a circuit consists of electrical wires. Electrical devices like appliances are one of the examples of this type of electricity.

Types of charges

Objects can exist in three different charged states. They are: positively charged, negatively charged, or neutral (no charge). Most objects are neutral. For example, we can say about the atomic particles in an atom of magnesium. Magnesium contains 12 protons (positively charged) and 12 electrons (negatively charged). An atom of magnesium is considered neutral since (+12) + (-12) = 0 (no charge). As soon as magnesium gains or loses electrons, it will have a charge on it. For example, magnesium will lose two electrons to become a stable atom so that it will have 12 protons (+12) and 10 electrons (-10). As a result, the overall charge on an atom of magnesium that has lost two electrons is (+12) + (-10) = +2. Therefore, magnesium is positively charged. Note that objects can only lose electrons. Atoms are not able to lose protons because they are held very tightly within the atom's nucleus. When two objects are rubbed together, electrons leave one object and the other object gain those electrons. As a result, every time objects are rubbed together, one becomes positive and the other becomes negative. When one loses electrons and the other one gains electrons, this thing is happened.

The law of electric charges

The Law of Electric Charges states that objects with different charges will always be attracted (move towards) each other. On the other hand, objectives with the same charge (for example, two negatively charged objects) will always repel away from each other. Neutral objects will always be attracted to positively or negatively charged objects as well. Neutral objects will never be attracted or repelled from other neutral objects.

Current Electricity and Static Electricity

When electron flows, current electricity is generated. Current electricity can travel along a path or a circuit from where the electricity is generated. Hydroelectric power plant is one of this example. In this case, waterfalls through turbines and power generators electricity. When an object becomes charged by gaining or losing electrons from rubbing against another object, then it is called static electricity. Static electricity can't move.

CHAPTER-20
HOW DOES A SOLAR THERMAL POWER PLANT WORK?

A power plant which collects and concentrates sunlight for harvesting solar energy required for heating up a fluid to a high temperature in order to generate electricity is called solar thermal power plant. Heats are transferred from fluid to water and creates extremely-heated steam. Turbines in the power plant can be rotated by utilizing this steam. A generator is used in order to convert this mechanical energy into electricity. In conventional electricity generation system, combustion of fossil fuels is required which causes environmental pollution. But, in solar thermal power plant, only heated steam produced from solar energy is used to generate electricity. In this type of power plant, solar collectors are used to concentrate the sun's rays on one point for getting required high temperatures. In order to collect solar radiation and store it, two types of systems are used. They are: passive systems and active systems. Solar thermal power plants are considered active systems. These plants are designed in such a system that it can operate by using only solar energy.

Types of Solar Thermal Power Plants

Although there are various types of solar thermal power plants, all of them work on same principle wherein mirrors are utilized in order to reflect and concentrate sunlight on a point. Solar energy is collected and converted to heat energy at this particular point. It produces extremely heated steam which helps to rotate turbines and run the generator in order to generate electricity.

Parabolic Troughs

Parabolic Troughs are also familiar as line focus collectors. They consist a long, parabolic shaped reflector that concentrates incident sunlight on a pipe that runs down the trough. A single-axis solar tracking system can be used in collectors in order to get maximum solar energy incident on the mirrors. When the trough focuses sun at highest times its normal intensity, then the receiver pipe in the center can reach highest temperatures. These troughs are lined up in rows on a solar field. When it is run through the pipes in the parabolic trough, a heat transfer fluid is heated. This fluid then returns to heat ex-changers at a central location where the heat is transferred to water, generating high-pressure extremely heated steam. This steam helps to rotate turbines to run generators and generate electricity. After then, the heat transfer fluid is cooled and return back through the solar field.

Parabolic Dishes

In this types of dishes, motors are used to track the sun. These are large parabolic dishes. In this case, highest possible amount of incoming solar radiation is received in order to concentrate at the focal point of the dish. These dishes are comparatively better than parabolic troughs in terms of concentrating sunlight. They can concentrate maximum sunlight and the fluid flows through them can achieve highest temperatures. In order to covert heat to mechanical energy, a Stirling engine is used in this system. It is done by compressing working fluid when cold and letting the heated fluid to flow outward in a piston or move through a turbine. After then, this mechanical energy is converted to electricity by the generator.

Solar Towers

These are large towers which perform as a central receiver for solar energy. These solar towers are located at the center of a large array of mirrors. They concentrate sunlight on a particular point in the tower. The heat exchange fluid is warmed in a heat ex-changer which is mounted in the tower. In this particular point, the heat can be extremely concentrated than incident sunlight. Steam is produced by the extremely hot fluid which assists to rotate turbines and run generator in order to generate electricity.

CHAPTER-21
WHAT IS GREEN HYDROGEN AND HOW IS IT PRODUCED?

Hydrogen is arising as one of the main choices for putting away energy from renewables with hydrogen-based fills possibly shipping energy from renewables over significant distances - from districts with bountiful energy assets, too eager for energy regions great many kilometers away. You could experience the terms 'dim', 'blue', 'green' being related while depicting hydrogen innovations. Everything boils down to how it is delivered. Hydrogen transmits possibly water when consumed however making it very well may be carbon concentrated. Contingent upon creation techniques, hydrogen can be dark, blue or green - and now and again even pink, yellow or turquoise. Be that as it may, green hydrogen is the main sort created in an environment unbiased way making it basic to arrive at net zero by 2050.

What is green hydrogen? How can it vary from conventional discharges serious 'dim' hydrogen and blue hydrogen?

Hydrogen is the easiest and littlest component in the intermittent table. Regardless of the way things are created, it winds up with a similar without carbon atom. Notwithstanding, the pathways to deliver it are exceptionally different, as are the outflows of ozone depleting substances like carbon dioxide (CO_2) and methane (CH_4).

Green hydrogen is characterized as hydrogen created by parting water into hydrogen and oxygen utilizing inexhaustible power. This is a totally different pathway contrasted with both dim and blue.

Dark hydrogen is customarily created from methane (CH_4), split with steam into CO_2 - the fundamental offender for environmental change - and H_2, hydrogen. Dim hydrogen has progressively been delivered likewise from coal, with fundamentally higher CO_2 emanations per unit of hydrogen created, such a lot of that is much of the time called brown or dark hydrogen rather than dim. It has no energy progress esteem, a remarkable inverse.

Blue hydrogen follows a similar interaction as dim, with the extra advancements important to catch the CO2 created when hydrogen is parted from methane (or from coal) and store it for long haul. It isn't one tone yet rather an extremely wide degree, as not 100 percent of the CO2 created can be caught, and not all method for putting away it are similarly compelling in the long haul. The primary concern is that catching enormous piece of the CO2, the environment effect of hydrogen creation can be diminished altogether.

There are innovations (for example methane pyrolysis) that hold a guarantee for high catch rates (90-95%) and successful long-term stockpiling of the CO2 in strong structure, possibly such a ton better than blue that they merit their own variety in the "hydrogen scientific classification rainbow", turquoise hydrogen. In any case, methane pyrolysis is currently at pilot stage, while green hydrogen is quickly increasing in light of two key advances - sustainable power (specifically from sun based PV and twist, however not just) and electrolysis.

Dissimilar to inexhaustible power, which is the least expensive wellspring of power in many nations and locale today, electrolysis for green hydrogen creation necessities to increase and lessen its expense by no less than multiple times over the course of the following ten years or two altogether. Electrolysis is economically accessible today and can be obtained from different global providers at the present time.

What are the benefits of energy change arrangements towards a 'green' hydrogen economy? How is it that we could progress to a green hydrogen economy from where we are at present with dim hydrogen?

Green hydrogen is a significant piece of the energy progress. It isn't the following quick advance, as we first need to additionally speed up the sending of inexhaustible power to de-carbonize existing power frameworks, speed up jolt of the energy area to use minimal expense sustainable power, before at long last de-carbonize areas that are challenging to charge - like weighty industry, transportation and flying - through green hydrogen.

It is vital to take note of that today we produce huge measure of dark hydrogen, with high CO2 (and methane) emanations: need is begin de-carbonizing existing hydrogen interest, for instance by supplanting smelling salts from flammable gas with green alkali. Late investigations have ignited a discussion about the idea of blue hydrogen as a change fuel till green hydrogen becomes cost-cutthroat.

How might green hydrogen become cost cutthroat versus blue hydrogen? What kind of essential speculations need to happen in the innovation advancement process?

The initial step is to give a sign to blue hydrogen to supplant dark, as without a cost for transmitting CO2, there is no business case for organizations to put resources into complicated and exorbitant carbon catch framework (CCS) and geographical stockpiles of CO2. When the structure is to such an extent that low-carbon hydrogen (blue, green, turquoise) is serious with dim hydrogen, then, at that point, the question arises whether it's a good idea to consider CCS as abandoned resources and when green hydrogen become less expensive than blue hydrogen.

The response will obviously contrast contingent upon the locale. In a net zero world, an objective that an ever increasing number of nations are focusing on, the excess outflows from blue hydrogen would need to be counterbalanced with negative discharges. This will include some significant pitfalls. In equal, gas costs have been extremely unstable recently, leaving blue hydrogen cost exceptionally associated to gas cost, and presented not exclusively to CO_2 cost vulnerability, yet additionally to flammable gas cost unpredictability.

How can green hydrogen be utilized?

Green hydrogen can be utilized as follows:

- Energy component hydrogen electric vehicles and truck.
- Compartment ships fueled by fluid smelling salts produced using hydrogen.
- "Green Steel" processing plants consuming hydrogen as a hotness source as opposed to coal.
- Hydrogen-controlled power turbines that can produce power on occasion of pinnacle interest to assist with firming the power framework.
- As a substitute for gaseous petrol for cooking and warming in homes.

For green hydrogen, notwithstanding, we could observer a comparable story to that of sun based PV. It is capital concentrated, hence we really want to lessen speculation cost as well as the expense of venture, through increasing assembling of sustainable advances and electrolysers, while making a generally safe off-take to diminish the expense of capital for green hydrogen speculations. This will prompt a steady, diminishing expense of green hydrogen, rather than an unpredictable and possibly inflating cost of blue hydrogen.

Sustainable power advances arrived at a degree of development as of now today that permits serious inexhaustible power age from one side of the planet to the other, an essential for cutthroat green hydrogen creation. Electrolysers however are as yet sent at tiny scope, requiring a scale up of three significant degrees in the following thirty years to lessen their expense triple.

www.ingramcontent.com/pod-product-compliance
Lightning Source LLC
Chambersburg PA
CBHW060436220526
45465CB00008B/3159